# Didactic Essays

From a piece of dark matter, somewhere in the Milky Way?

## Damon Dion Reed

authorHOUSE®

*AuthorHouse*™
*1663 Liberty Drive*
*Bloomington, IN 47403*
*www.authorhouse.com*
*Phone: 1-800-839-8640*

*© 2012 Damon Dion Reed. All rights reserved.*

*No part of this book may be reproduced, stored in
a retrieval system, or transmitted by any means
without the written permission of the author.*

*Published by AuthorHouse 10/4/12*

*ISBN: 978-1-4772-7699-0 (sc)*
*ISBN: 978-1-4772-7695-2 (e)*

*Library of Congress Control Number: 2012918664*

*Any people depicted in stock imagery provided by Thinkstock are models,
and such images are being used for illustrative purposes only.
Certain stock imagery © Thinkstock.*

*This book is printed on acid-free paper.*

*Because of the dynamic nature of the Internet, any web addresses or
links contained in this book may have changed since publication and
may no longer be valid. The views expressed in this work are solely those
of the author and do not necessarily reflect the views of the publisher,
and the publisher hereby disclaims any responsibility for them.*

# Dedication

I would like to dedicate these symbols, words, sentences, punctuation, paragraphs, and chapters to the mental exploration of science. Only with science, will we be able to achieve harmony.

# Contents

| | |
|---|---|
| Essay # 1: Shhhhhhhhhh | 1 |
| Essay #2: Song : Sentence :: _____ : _____ | 6 |
| Essay #3: Talking to Books | 9 |
| Essay #4: Beyond Reason | 11 |
| Essay #5: The Benjamin Button Theory | 14 |
| Essay #6: Piece of the Puzzle | 24 |
| Essay #7: Gravity | 32 |
| Essay #8: Hysterical Electrons | 49 |
| Essay #9: Quanta Dynamics | 56 |
| Essay #10: The Sparkle of Darkness | 59 |
| Essay #11: The Tower Perspective | 61 |
| Essay #12: Do NOT Tilt! | 63 |
| New Age Adages and Addendums | 65 |
| Glossary | 81 |
| References | 85 |

# Essay # 1: Shhhhhhhhhh

"Aaaaaaamen good Lordie. Aaaaaaaamen have mercy." Oh, hi there. You caught me in the middle of one of those singing thoughts. How embarrassing. If you could see me right now, I'm turning bright red. I know this is only my third scientific book, but I should have known better than to start a book without a coherent thought, thesis, or non-thesis.

Whammy, the beginning of my book! I know you're impressed because I'm COMPLETELY shocked. Mostly because I have no clue as to where I'm going with all these words: Words to fill up pages, pages to fill up books, and books to fill up my shelves. WAIT. I don't have any shelves and I think that was my first micro-thesis: This is not TV because there is an abundance of words. Miniature brain fart on the micro-thesis? And now, a new paragraph.

I'm so excited. I'm on the third paragraph of my book and things are going quite dashingly, perhaps a little wordy, but mildly entertaining none-the-less. So where do I want to go with this paragraph, chapter, and book? Think Winnie the Pooh…Think! Oh yes, science. I want to write a book about science, but I have all these tangential thoughts arcing out in every direction. WAIT. I'll call these words a collection of essays, thus hiding my absent mindedness. But, I also want it to sound intelligent to the archaeologists that might find this book. Ah ha, big words! Big words sound intelligent... and confuse people. Ummmm, maybe that isn't such a good idea. Oh hell, I only write my third scientific book once. Anyway, isn't that what a glossary is for? On second thought, scratch that idea. It would be too much work to put in a meaningful glossary.

I will call this book: <u>Didactic</u> Essays from a piece of dark matter somewhere in the Milky Way. Wait, that is in the past because I already titled this book as such. Déjà vu. Pardon my French. It is almost as if one has to write a book to come up with the title. Thankfully, all that is in the past...I think. Let's look towards the future and the words that I'll have to think up, jumble together, and punctuate improperly. Words that I'll have to define, redefine, and make rhyme. NOT.

Okay. Paragraph four. I know I'm about to say something intelligent, something profound. I mean, I did use the word <u>didactic</u>, which means scholarly. And, I alluded to science and dark matter...Ooooooo Mysterious! Something intelligent is on the tip of my tongue. Tip of my tongue? What a silly <u>adage</u>. Boo-yah, big word number two! I'm on a role now. Hold on a second. If you're thinking that adage means proverb, then I probably sound like a bone-head. But let me assure you that I looked up the word <u>saying</u> in the *Oxford Dictionary* and <u>adage</u> was a synonym. And you know what else? I'm not going to reference it because definitions change, all dictionaries are NOT the same, and the people who pick the words for the Graduate Record Examination (GRE) need some tranquilizers. More specifically, the GRE people need some anal-suppository tranquilizers. And with that, I bring paragraph number four to a gentle, butt roaring close. (FYI, that was a <u>malapropism</u>.)

*Didactic Essays*

You know how I said, "Tip of my tongue?" Well, what I meant to say is this: "Tip of my neural synapse." As to which neural synapsis, I can't remember. But wait, how am I supposed to remember which neurotransmitter I used if I'm supposed to come up with something new/unique/original? Where do those neurotransmitters come from and what-the-hell is a synapse? Ha, I tricked you! Yeah me! I mean, sorry. You probably thought there wasn't any intelligence left in me, which is probably true. In any event, your mother and I have noticed that you've been using your brain and we think it is time we had The Talk. LOL, I'm sorry. If you're reading for continuity, then the statement "your mother and I" made very little sense. I just thought it sounded funny.

The Talk: There is this gray-mushy thing between your ears that squishes out stuff (neurotransmitters) into certain places (synapses) where other things (nerve cells) suck them up faster than you can blink an eye. Not only does your brain squish and squirt, but what you EAT plays a big part in how your brain squishes and squirts. Here is an adage for all of that: You are what you eat. WAIT. That doesn't really fit the paragraph. How about this: What you EAT affects how you think? I know that it's NOT as catchy, but intelligence doesn't seem to be <u>contagious</u> (GRE word #4).

"On the road again, I can't wait to be on the road again!" Lyrics by a really famous dude, but I can't remember which synapse I stored his name.

I'm so happy to be back in the saddle slinging words to the right, down 1.5 spaces, and then to the left. It's like a dream come true, except I'm completely <u>cowed</u> by <u>zealots</u> (GRE words 5 & 6). So be weary my friends. The first rule of Word Club is: Visualize Peas. The second rule of Word Club is: Visualize Peas. The third rule of Word Club is: Don't let the cows out! Apparently it takes a long time for the cows to come home because someone made an adage for it: When the cows come home. Ok, mental song break.

"Who let the cows out…Moo, moo, moo! Who let the cows out?" Wait, am I the only one who sings this song?

All right. Before I started this paragraph, I <u>chafed</u> my hands together, blew on my fingertips, and said a little prayer. I can feel those neurotransmitters a squishing and a squirting. I'm on the brink of an idea, but I don't know where to start. I mean, I want to write a book about science with a ton of big words, but I don't want it to be <u>arduous</u> (GRE word #9). I guess I'll just start with some big words and go from there.

But wait! Don't flip to the next essay just yet. I have more words for you to read, absorb, and then say to yourself: "Ok, this **word** thing is getting annoying."

I completely agree. It's fucking <u>invidious</u>! (GRE word #10 is <u>invidious</u>, not fucking.) I'm sorry for NOT using a <u>euphemism</u> (GRE word #11), but I really want you to stop reading (NOT YET!), flip to the glossary (Just kidding, that won't help you.) and READ the definition of <u>invidious</u>. Ok go, I'll wait. (Insert Jeopardy music here.) Are you back yet? Hello? I said I was only kidding. There's NO meaningful glossary! Oh, there you are. So what do you think? Am I right-on or what? The words we use to <u>communicate</u> (Not a GRE word.) can cause us to build walls <u>metaphorically</u> and physically.

"Gorbachev, tear down this wall" - and we'll make millions selling little pieces to tourists. Okey-dokey, now I have to apologize to all the Reagan fans. Sorry?

Do you see the problem here? (Yeah, I know there wasn't a

*Didactic Essays*

transition from the last paragraph, but do we always need such <u>frivolous</u> words between thoughts?) The words we use to communicate can be defined one way or another, redefined, interpreted, and used as motor to build walls, which can eventually be torn down and sold for pennies on the dollar.

Here's another adage: It's all about the money.

Shhhhhh! Before you say anything, think about the words you use! (BTW, that was a <u>cant</u>, which is GRE word #n+1)

# Essay #2: Song : Sentence ::

# _ _ _ _ _ _ : _ _ _ _ _ _

Germinate…Gerontocracy…Gerrymander…Oh hi there. If you haven't noticed, I've been studying for the GRE because it is one heck of a test. Did you know that <u>factitious</u> means artificial or sham? Yeah, that one was a shocker to me too? If you're wondering why the preceding sentence had a question mark, it was because I felt like it? Anyway, another thing about words is that they are often <u>factious</u> or cause dissension. But none of this really matters because we just want to learn enough to pass the test, get the grade, get the job, get a beer, get a promotion, get a vacation, and get laid…right?

On the GRE there are these really cool questions called analogies, which is comparable to the GRE word <u>analogous</u>. First, the people who wrote the test give you one word that can be defined a couple different ways. Then you have to find out how the first word correlates to the second word, which can also be defined a couple different ways. Finally, you're supposed to pick the pair of words that have the same type of correlation when you USE the test-maker's definitions. Needless to say, I get word induced <u>aphasia</u> when I look at them. Therefore, I made up an analogy to practice. By the way, the analogy

*Didactic Essays*

questions are read as such: Song is to Sentence as _____ is to _____?

Song : Sentence :: _____ : _____

- A) <u>Ennui</u> : <u>Gregarious</u>
- B) <u>Febrile</u> : <u>Herpetologist</u>
- C) <u>Fatuous</u> : <u>Jingoist</u>
- D) <u>Droll</u> : <u>Sanctimonious</u>
- E) Turd-burglar : <u>Neologism</u>

Let me <u>delineate</u> you a picture. If you think <u>delineate</u> means portray, then the preceding sentence made a little sense. If you think <u>delineate</u> means depict, then you get a slight <u>descry</u> of what the sentence is trying to depict. And finally, if you think <u>delineate</u> means sketch, then you get into graduate school **if and only if** you can correctly guess the meaning of the other words, their correlation, and eliminate the inappropriate choices based on their variable definitions. It's nothing to <u>deride</u> about, especially when you have twenty-nine other questions to answer in thirty minutes. Ok, mental song-break.

"Automatic Booty. Zero to tutee fruity!" Beck song quote.

Are you better now? Me too? But, I wonder who gets the hard or easy tests. I mean, you have to pre-order the GRE, give a butt-cheek impression, and submit a written <u>affidavit</u> stating that you've never <u>nettled</u> anyone in your life. All of which, means someone somewhere is deciding who gets the hard or easy test. Maybe the people with the second letter of their middle-last name between 'A' and 'C' get the easy test. In which event, <u>DUH</u>! (DUH is not a GRE word. I just like underlining things.)

If you're still wondering what the answer was to my practice analogy, wait no longer. The answer is simple. If you know that <u>ennui</u> means boredom, <u>droll</u> means amusing, <u>febrile</u> means feverish, <u>fatuous</u> means brainless, <u>gregarious</u> means sociable, <u>herpetologist</u> means the study of reptiles, <u>jingoist</u> means a war-like chauvinist, <u>sanctimonious</u> means devoutness, and <u>neologism</u> means a newly coined word, then Song is to Sentence as <u>Jingoist</u> is <u>Fatuous</u>. (FYI, the previous words were used in the same order to define the same symbols arranged in the same order in the GRE book I was reading, except for turd-burglar. That was a <u>neologism</u>? Please draw the appropriate correlation to the GRE book referenced later in this book.)

Are you <u>obfuscated</u>? I know that my logic may be a little <u>oblique</u>, but any <u>ostentatious</u> <u>ornithologist</u> can see that I'm just trying to learn new words by writing Mad Libs! By the way, did you know that <u>Lib</u> is colloquial for liberation and <u>Mad</u> is colloquial for silly? Silly Liberation?

Now that I have all that silly liberation out of the way, enough big words to appease any <u>intellectual</u> (Sadly not a GRE word.), and have completely annoyed everyone else, I think it is time for some <u>ratiocination</u>. If you're not annoyed by all these big words or don't consider yourself an intellectual, then you should give yourself an intellectual promotion for waging through them. If you are an intellectual and are <u>vexed</u> by such moderately priced intellectualism, then <u>rebuke</u> yourself and go to the corner. As for me, my <u>repartee</u> is this: Why can songs mutter words that sentences dare not contain? Are songs superior to sentences? Or, do sentences lack good <u>oscillations</u>? And finally, who sings that song?

# Essay #3: Talking to Books

Talking to books used to be one of my idiosyncrasies; however now I just think at books and everyone seems happier now… except for the books. I don't know how they feel. I guess it is better than laughing at a book. None-the-less, how does one write <u>bombastically</u>? Am I doing it **write** now? Was that a <u>malapropism</u>? Nope, I mix those words up all the time. And here's one final note before I actually start this essay:

"To whom it may concern,

On page 189 of Barrons 2008 GRE test preparatory book, there is a definition missing. For a moment, I thought <u>martinet</u> meant "no talking at meals." Thankfully, I looked it up elsewhere and it means "strict disciplinarian." Please reference the reference I just referenced and make sure to correct it on all subsequent editions.

Sincerely yours,

Someone who is about to <u>masticate</u>, my lunch of course."

Okay, where were we before that <u>ungainly</u> <u>verboseness</u> interrupted our conversation? Ah yes, words. I was trying to be <u>urbane</u> with words in order to venture toward the fundamental question: Are we really any different? Aside from the <u>minuscule</u> <u>anomalies</u> in our <u>pensive</u> <u>genetic</u> <u>genealogy</u>, we're the <u>propellant</u> of the future, which is now the past… or something like that. Let's not be so <u>punctilious</u>.

"The positron is connected to the proton, which is connected to the neutron, which is in the atomic nucleus." Scientific song break.

I've decided to sing all my scientific postulates because they are too bothersome in the written form. In the written form, they exist in the non-existent category of scientific satire. Therefore, I'll just going to sing my scientific postulates instead. How about this:

"One little, two little, three little positrons!" Scientific-satire song break.

Damn it, that doesn't have a good ring to it. Now I'm at a loss. I want to educate; however nobody wants to read about science anymore, and I have no musical talent. I tried the guitar, but I lack that ability. I pound on the piano sometimes, but that isn't music. I guess the only thing I have left is words. Words that I type, send out into the world, and hope that people don't take them the wrong way, get angry at me, each other, or anyone else.

I want to write <u>words</u> that make people smile. I want to write <u>words</u> that make people feel happy. I want to write <u>words</u> that bring us together and not tear us apart. But enough with all that mushy sentimental stuff. Let's talk about something else.

# Essay #4: Beyond Reason

I finally took the GRE…Whew! I blew chunks on the verbal section! All of which, means that I'm not trying to learn anymore big words…Yeah! So how are you-all doing? I've been so busy preparing for the GRE that I forgot about you and <u>science</u>. Actually, I was going to write a whole essay about desensitization, American media, and how we all ignore the world to some extent, but I decided against it. Actually the more I think of it, the less motivated I am to write. I mean, what good is another string of verbs and nouns in comparison to what you were taught as a kid? It is not like you're going to jump out of your seat and say:

"I'm going to change my life! Where's the closest vending machine!"

On the off chance you do get that feeling, could you please get me a Hershey's candy bar?

Now that I've started another essay with nothing to say, I guess I should get busy saying it. Actually, I wrote a short book titled *Ham Semantics*, but now I don't feel like sharing it. I mean, after pondering the contents a little more, the propagation of thought along the venue of quantum mechanics will simply result in lasers that can destroy protons, which is NOT what any of us want? So, I'm stuck pondering things that are beyond reason, which isn't reasonable. Therefore, I'm going to study something useful until I find a reason worth writing about.

Actually, scratch that idea. I just remembered what I wanted to say: Daaaaaaark Maaaaaatter! Dark matter is that scary stuff that exists in **every other part** of the universe but NOT in **this** solar system. WAIT. How is that possible? Did we scare all the dark matter out of **this** solar system? I'm so confused. Excuse me while I UN-confuse myself. Actually, that made everything more confusing.

So there is this stuff called dark matter that **can't** be seen and exists in **EVERY** solar system except this one. At least that is what the theorists say. But this is what I say: "Bad theorist, bad theorist!" Now that I'm back on track, relatively speaking, let's talk about **DARK MATTER**. You see, no matter how wonderful the idea of dark matter might be in explaining away theoretical inconsistencies, it is highly UN-likely that our solar system forgot to pick up dark matter at the galactic grocery store: Dark Matter isle seven next to the string.

I know that most of you are probably scratching your noggin, but there is no need. It is quite simple. If we define dark matter as **MATTER** that does NOT release light, then take one step to the left, one step to the right, and then repeat. Now a piece of dark matter is doing the cha-cha! Yes, you've got it! Hold that thought...hold it...**You're a piece of dark matter.** Also, that was a pop-culture reference to the movie entitled *Real Genius*.

Now I'm sure that some smart-bootie will say that everyone is releasing infrared radiation as well as butt sound waves, but let's be honest. Unless there are Navy SEALS with very powerful night-vision goggles in our back yards, then we are all considered as galactic dark matter. Hold on a second. Okay, I don't see any Navy SEALS in my back yard. Navy SEALS, Navy SEALS, ah right there. Sorry, I had to check "Navy SEALS in my back yard" off my list of things that would freak me out. Not because Navy SEALS are freaky, but because their existence in my back yard would be freaky.

With all that said, let's have a nice big pity party! All those people who were able to sleep at night because there was NO dark matter in OUR solar system, are going to FREAK OUT!

"Ahhh! The sheets are dark matter! Ahhh! The pillows are dark matter! Ahhh! That person sleeping next to me is dark matter! Oh the horror, the HORROR!" Freaked out person.

Should someone get them a glow in the dark tissue? Wait, would things be a whole-lot weirder if everything was completely luminous?

Well, that was an interesting essay. I'm not sure what there is to talk about, let alone if there is anything at all to talk about. Oh yeah, one other word I wanted to use was SPOONERISM. Unfortunately, I couldn't think of a good way to be humorous about it.

# Essay #5: The Benjamin Button Theory

Now that I've cleared up some silliness with regards to Dark Matter, it is time to move onto the theory of the Big Bang. Granted, I've already written a lot about the Big Bang, but this time I'm going to be a lot more specific. In order to do that, this essay is going to get a little hairy. By that, I mean it AIN'T going to flow like normal essays. You see, in order to understand the current theories of the universe, we need to understand math. And if you've ever read anything I've ever written, you'll already know that I have a problem with math. Luckily, I know a little bit about the creator of modern math, namely: Sir Isaac Newton.

Alas, I knew Newton well. He was a man of infinite jest… NOT! Newton was neurotic, antisocial, and a mathematical genius. Unfortunately, Newton thought that the universe had limits. Literally, he thought there were regions of God and then there were regions of the Devil, or something like that. Actually, I have no clue why Newton thought the universe was an **energetically** CLOSED system. Maybe it made his calculations a little more accurate, precise, or meaningful? In any event, I'm going to dig up Newton's remains and kick him in the nuts by saying:

*Didactic Essays*

"What, God never lets ENERGY disappear but allows people to?"

Here is the gist of all that silliness: If the universe is a CLOSED system where energy can NEVER be lost as the universe expands into new regions of space, then the mathematicians have created a cyclic fairytale that can never be questioned. Hopefully, the LOW-GIC is clear: If the universe is expanding and energy is being used to heat up regions of space at absolute zero, then those regions of space that are NOW above absolute zero will **NEVER** return to absolute zero because of mathematical magical energy that can never disappear? Nooooooo, THAT doesn't make sense! My understanding of the universe, math, and God is now in question. I can't take these inconsistencies in SCIENCE! Quick, someone paint over these ERRORS... Oh the humanity! Wait, science has nothing to do with humanity.

This mathematically CLOSED system is the whole reason why scientists postulated the Benjamin Button Theory. Actually, it isn't called the Benjamin Button Theory, but it is like the movie *The Curious Case of Benjamin Button* with Brad Pitt and Cate Blanchett. WAIT. Scientists don't understand pop-culture references. Anyway, as a result of the universe supposedly being a closed system where energy NEVER disappears, scientists postulated that the universe will stop expanding and start contracting. All of that is a fun to think about, but there is one problem: Entropy. Now if you don't know about entropy, then here is the gist of that: Entropy is the concept that energy LIKES to diffuse. For example, stop heating water and it will cool. Unfortunately, as a result of the Benjamin Button Theory, scientists have postulated that once all the energy in the water has gone away, the water will begin to heat up again...or something like. Actually, that was a really horrible example because there are NO REAL examples that are comparable to what the scientists have postulated with regards to the universe. In short, scientists think that the universe will create ANTI-entropy at some point, which isn't really much of a theoretical jump because they've already postulated **exactly** when gravity was birthed. For those of you who don't know when that was, gravity was born sometime after entropy… relatively speaking, of course.

Unfortunately, as a result of HOW modern science seems to work, this is probably the rationale someone gave a long-long time ago: If there is ANTI-matter, then there must be ANTI-entropy. My reply to all this nonsense is this: "Please, as if!" Just because there is a **NAMED** type of matter that releases ANTI-matter when combined with another NAMED type of matter, that doesn't mean that there is ANTI-entropy. That would be like me saying:

"Since sodium hydroxide combines with a proton to make water, then there is ANTI-water."

Honestly, what is so difficult to comprehend about the NON-existence of ANTI-entropy? What, has the simple concept of nomenclature turned us all into illogical tape-recorders? Thankfully, I've just received word that Mr. Entropy is going to retire. Here's his letter of resignation:

> To whom it may concern,
>
> For over a hundred years, I have given science my whole hearted attention. I have absorbed the short comings of my fellow theories and reluctantly given my approval to be held responsible for things and events for which I am NOT responsible. I have been stretched so far that my fellow theories refer to me

*Didactic Essays*

as: The HOLEY-grail of Knowledge. If energy is being lost, released, or transformed, everybody turns to me. I no longer can stand for this tarnishing of my good name. Therefore, please accept this letter as my resignation from the position of executive director of the universe.

Kind Regards,

Mr. Entropy N. Zane

The mathematical apocalypse has begun. Rise up theoretical brothers, choose a side, arm yourself with an equation, and tape your glasses with camouflage duct tape. Mr. Entropy has retired and there will be a war. Then there will be a holy war, a poor war, and calculator war, which is just a race to see who can spell BOOB first.

If Mr. Entropy is no longer the director of the universe, then we need to find someone else to blame for all OUR mistakes. We cannot go through the day or night without knowing the fundamental sequestering nature of nature, or the universe. So rise up, grab your theoretical pens, your shielding calculators, and favorite stress-releasing hand squishy toy. Disregard your wife, forget to feed your cats, and barricade yourself in the spare bedroom with all those old physic books, which are only, at best, relative to Earth. On second thought, maybe we should just pull a *Weekend at Bernies* with Mr. Entropy after he kicks the bucket? Ooooops, there's another pop-culture reference. Sorry. Sometimes I get the distinct feeling that scientists are not part of pop-culture.

After reading about Mr. Entropy's retirement, the pertinent question should be this: Where should we go from here?

We could either throw Mr. Entropy a retirement party with those little pointy hats, or we can convince him to stay on as a consultant. Either way, there are some things we need to realize.

The universe is defined as: Regions of space that are NOT at absolute zero and A LOT of space at absolute zero. Unfortunately, no one has realized that warm regions of space can cool back down to space at absolute zero, which is completely MATH's fault. A mathematically closed system has resulted in people postulating magical energetic quanta that can never disappear. Fortunately for Newton, this conspiracy goes even deeper.

It has been said that math is the language of the God(s), depending if you're a mono or poly-theistic. Unfortunately, within the framework of Earth based logic, this statement is incorrect. The Creator(s) probably didn't sit down and write a very long equation to create and dictate the universe. The Creator(s) probably just created the universe, and the repetitive properties within the universe gave birth to mathematics.

If the universe was created **without** math in mind, then math is simply the expression of the repetitive events in the universe. This means that the use of math as a guide to create an extrapolative-theory of the universe will result in a tangential theory that leaves all **rationality** behind as it streaks out into the **conceptual** void. In short, if NO equation can contain enough variables for every single aspect within a universal system, then it is highly unlikely that an extrapolative math based theory of the universe will be helpful in understanding the universe. That is not to say that math isn't great at describing the repetitive events within a very small portion of the universe, Earth,

*Didactic Essays*

for example, but the use of math to extrapolate a universal theory will be counter-intuitive. Therefore, we need to use more fundamental laws to create a universal theory. For example, how about these?

1. Entropy causes energy to diffuse.
2. Opposite charge attraction (+/-) battles Entropy.

Unfortunately, things get a little more complicated with the introduction of opposite charge attraction. You see, Math has its hands in this arena as well. Therefore, let's start with the illogical theory of gravity and work our way towards the Math kerfuffle. To do this, here is a *Dreamstar* original play titled *The Power of the Radius*:

Act I

The curtain lifts to absolutely nothing. In fact, there is such an absence of anything, including energy, that this NOTHINGNESS is really-really COLD: Absolute zero to be empirically precise and accurate. Then, out of nowhere… POP! A star comes into existence and it is really-really HOT. Nobody knows if stars represent absolute HOTNESS, but to the audience, it appears to be the absolute contrast to absolute zero. Therefore, with the two main characters introduced to the audience, absolute HOTNESS and absolute ZERO, the curtain lowers to a bone-chilling scream,

"What is Gravity?!"

Act II

The curtain rises to introduce another character: A second star adjacent to the first star. The two stars are not touching, but it is as if they are drawn together because of their really

HOT salsa dance. In every direction, coldness surrounds these two stars, except for the space between these two stars. Be it mere happen stance that the space between these two stars is LESS harsh than the absolute zero space everywhere else, it matters NOT. In a dark and cold universe, these two fiery stars are being pulled together as if by love. In fact, these two stars are so close together that their dance is one of ill-fated destruction, but beautiful none-the-less.

Act III

As the curtain draws away yet again, a place that is neither absolutely hot nor cold comes into view. It is a place where the stuff of stars has amassed into spheres. It is here that an equation that describes charged specs of stardust, is ALMOST exactly the same as the equation that describes the stars' HOT salsa dance. One equation denotes the attraction of opposite charges as a superfluous force and the other denotes the force as gravity. Slowly, the curtain falls on this tragedy of coincidence.

(The curtain closes to applause, howls, and sniffles?)

"Doesn't a really good tragedy bring to heart all the emotions that define us as a species? And that Reese Witherspoon and George Clooney were absolutely spectacular in Act II. Their tantric salsa dance was a little over the top, but good none the less." Theoretical Theater goer.

As you might be able to ascertain from this play, the universe is a cold harsh place. Also, the diffusion of thermal energy might not be as powerful as the attraction of opposite charges, but every amount of energy consumption helps. Now, I bring all this up for a couple reasons. First, gravity does not have a thermal-FACTOR. Second, GRAVITY does NOT have a thermal-FACTOR? Wait, is someone pulling

*Didactic Essays*

my leg? How could anyone write a gravity equation with a UNIVERSAL-constant and forget to factor in temperature? Thirdly, why is the equation for opposite charge attraction similar to the gravity equation, which has a UNIVERSAL-constant? Maybe we should take a step back and think about all this?

So what do we know about gravity? Well, gravity is a NON-energetic **magical**-quantum that can traverse space WITHOUT losing any energy. Wait, do you smell that? Never mind. Since Newton and the apple, thousands of scientists have waged a war in the name of gravity. They have made an equation, determined a universal constant, and FORCED-science to come to a SCREECHING halt because the gravity EQUATION is able to describe a NON-energy-based FORCE in the entire universe? Maybe we should take another proverbial step back?

As a result of people just memorizing the gravity equation and UNIVERSAL-constant, nobody has given gravity any more thought. Here are some **problems** with gravity:

1. Gravity does NOT have a temperature factor. Maybe the mathematicians included a catch-all temperature factor in the UNIVERSAL CONSTANT, but how can someone factor in a range from zero to 100,000,000 degrees Kelvin? Seriously? Is the universal constant that good? Dam?

2. Gravity does NOT depend on the arrangement of matter/mass or the unique characteristics of elements. Really? Wait. So the density of the matter doesn't matter to gravity? Another Dam?

3. Gravity can traverse space at absolute-zero without losing ANY energy to space at absolute-zero? Ohm, Super Dam?

There you have it, boys and girls, GRAVITY: a NON-energy-based FORCE that rules the universe. Wait. I do not like where this is leading science. Unfortunately, no one is dumb enough to question centuries of science, a universal constant, or MATH…I I I should just give up?

Now to the untrained intellect, it might seem as if I've got a problem with math and its dirty-dirty word whore called FORCE, but math has been very contentious with me lately. It has been supporting arcing assumptions about the universe, creating asymptotic limitations, and allowing extreme wiggle room within science's extrapolative theories. I used to tolerate math's little vices, but it became so mentally abusive that I had to walk away from the relationship. I heard that math enrolled for rehabilitation at a derivative hospital, but we don't speak anymore. One thing I did love about math was its sexy asymptotes. Oh me, oh my! The way a region of a mathematical curve never reached a certain space on an X,Y coordinate system, really got my juices flowing. Theoretically, this is where I say something silly like this: "The mathematical curve for energy disappearance in a CLOSED system has an asymptote at absolute zero, which means that warm space can NEVER lose all of its energy because of a mathematical asymptote." Go MATH! If you couldn't see me when I wrote "Go MATH," I was doing my math cheerleader impression with one leg, arm, and calculator-pompom pointing towards the asymptotic regions of the universe?

I know that many of you feel uncomfortable knowing that **energy** can disappear in this universe, but don't worry so much. If you believe in the first law of thermodynamics, you can live in denial forever. Actually, that was a lie. You can live in denial for as long as your body survives.

**Figure 1: Thermodynamically Perfect House**

So what **should** you have learned from this essay? Denial ain't just a river, baby. It is a way of life. Also, there are many illogical things hiding in science, just waiting to be discovered. For example, you could have learned that the basis of the first law of thermodynamics is mathematics, which caused someone to postulate the Benjamin Button Theory of the universe. Or you could have learned that mathematicians have ascribed a UNIVERSAL constant to gravity that doesn't factor in temperature. Or you could have learned that there is a fierce-fierce battle between Entropy and energy conservation. Or you could have learned that matter was born with extra-baggage. Wait. That is the next essay. Never mind.

# Essay #6: Piece of the Puzzle

In all dishonesty, the saying "extra-baggage" was introduced by a clandestine-tribe of thermodynamicists in the early 1940's such that the Elite-Minds of Earth could agree about how the universe was formed, how galaxies were forged, and how stars existed energetically. (Footnote: Thermodynamicists isn't actually a word, but it's FUN imagining people trying to pronounce it.) Unfortunately, the term "extra-baggage" has been taken over by pop-culture and now nobody understands the Big Bang theory anymore. Damn you popular-culture! Anyway, let's reinvestigate the Benjamin Button Theory of the universe as it pertains to matter and see if we can find anything LOW-GICAL, which is just my code for illogical.

In the beginning, God created the Big Bang and the Big Bang created Entropy by exploding…or something like that. Then for whatever reason, Entropy begot Gravity, which collected all the matter that was **created** by the Big Bang to make stars. Then the stars, following the first law of thermodynamics, forged heavier elements without the SLIGHTEST degradation of the energy in matter. All of which means, the universe HAD to give birth to BABY-matter with an extra-bag full of energy to account for all the NON-matter-energy being released by stars. To complicate matters even more, gravity is a NON-energetic entity that

*Didactic Essays*

diffuses from large collections of BABY-matter, which means ALL the gravity in the universe was nicely packaged within BABY-matter's extra baggage. Fortunately, we didn't have to look very far to find something illogical. Therefore, let's try to find a more logical explanation for the arrangement of the universe.

Let's think about this. We know that the universe is a closed system. Wait. Let's postulate that the universe is **NOT** a closed system. If the universe is NOT a closed system, then energy can **degrade**. If energy can degrade, then the universe wouldn't have to make BABY-matter with extra-baggage. Unfortunately, this doesn't fix the gravity conundrum. Hmmmm, how about this: The universe contains a spectrum of energetic quanta categories that contain a spectrum of energetic quanta that are slowly degrading? If ALL energy is degrading as a result of Entropy, then the association of energy will prolong the degradation process. Fortunately, this leads to the following question: Is time directly proportional to space? By this I mean, is the rate of energy degradation directly proportional to the surrounding energy? If energy degradation is directly proportional to the surrounding energy, then there would be a hierarchy of energetic quanta. Kind of like the one I'm about to propose?

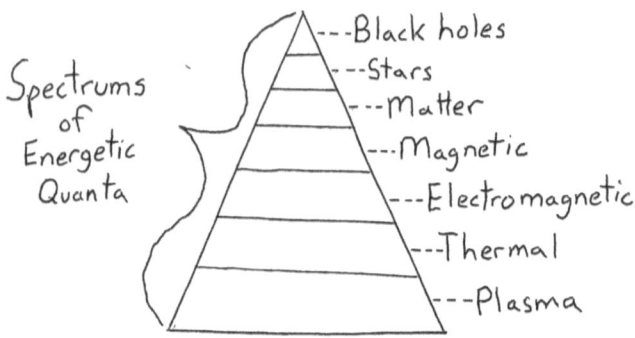

Figure 2: The Energy Pyramid's Extravaganza
of Energetic Quanta Spectrums (EQS)

We already know that electromagnetic radiation exists as a spectrum. Therefore, let's postulate that each category in the Energy Pyramid exists as a spectrum of energetic quanta. For example, you might not think that matter exists as a spectrum, but if you were to add in all the isotopes to the periodic table, then it starts to resemble a spectrum. Likewise, a bar magnet is NOT as strong as the Earth's core, which is a good indicator that magnetic energetic quanta probably exist on a spectrum. And finally, scientists have discovered that destroying matter in particle accelerators produces a spectrum of plasma, which I discussed at length in *Bibbles of Spuce-tame*. Therefore, let's take all this information and piece together a more functional theory… hopefully. Unfortunately to you, not me, I will first have to talk about Entropy again.

Hopefully by now, we have all pulled together a definition of Entropy, which probably goes something like this: Energy will diffuse unless there is a reason for energy NOT to diffuse. Here is an example for both scenarios: Turn a light on and then off. When you turn a light on, part of the

*Didactic Essays*

energy from matter, which is stable and usually doesn't diffuse quickly, is stimulated to release light, which diffuses. Granted, I could go into the extreme depth with regards to this phenomenon, but for the time being, let's just leave this example with a simple explanation: Degrading energy takes up more space than NON-degrading energy. In short, as the light diffuses out from the light source, the photons are taking up more space than if they were still incorporated in the matter.

For those of you who have been paying attention, that simple example just explained my postulate of "Highly Ordered Star Seeds." If you're not familiar with that postulate, then let me introduce it to you. In my book titled *Dreamstar*, I postulated that stars are Highly Ordered Entities. Granted, I also postulated that matter was created by the expansion of energy as a result of relative acceleration, but thankfully nobody read that mistake. Here is a **better** postulate: Because matter degrades into other energetic particles, it is very likely that Highly Ordered Star Seeds are degrading into matter. Actually, if we view the whole universe as a Systematic Battery system that is designed to inhibit the loss of energy, then this extrapolative postulate has even further ramifications. If the Highly Ordered Energy in a star is degrading into matter, then the Highly Ordered Energy in stars degraded from Highly Ordered Galactic seeds. The only question that remains after this is the following: How did the repetitive pattern of spiral galaxies propagate within the universe, and WHY?

If we think of a star as a battery system that is slowly degrading into matter, which is also acting like an insulator to the harsh universe, then how is a spiral galaxy mimicking a battery system? Well, in order to answer this, we need to talk about black holes. Even though many scientists have postulated that black holes destroy energy, I like to think that black holes are the universe's recycling bin. Energy goes in, and millions of years later, the black hole explodes to produce multiple smaller galaxies. Or put another way, black holes are the structural unit by which galaxies reproduce asexually into smaller self-replicating battery systems.

Imagine that there was one really big highly ordered galactic seed in the beginning. For simplicity sake, let's call it Galactic Seed-One (GS1). When GS1 degraded, it formed a circular galaxy with an empty center. The degraded "particles" of GS1 became stars, which degraded into matter. As these stars slowly degraded into matter while they moved in a circular pattern, the empty center of this galaxy became a very magnetically stressed region of space that recycled energy. As more and more energy was collected into this dark region, it was compacted and arranged to produce another Highly Ordered Galactic seed. In short, the arrangement of the GS2 is dictated by the movement of the degrading components of GS1, which are the stars of GS1. For example, when GS1 degraded, it formed a circular galaxy because there was no external energy to distort the movement within the galaxy. But, when GS2 degraded, it split up the primary galaxy into multiple smaller collections of stars, all of which formed similar circular or spiral galaxies.

*Didactic Essays*

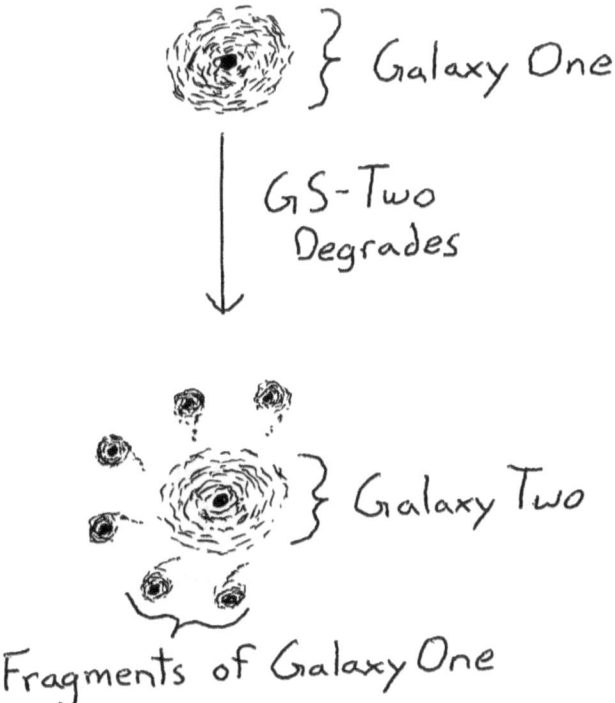

**Figure 3: GS-Two Degradation**

In this way, each galaxy repeated the cycle of star degradation into matter, matter destruction to form a new Highly Ordered Galactic seed, and subsequent asexual reproduction into multiple smaller galactic battery systems.

Now I'll be honest with you all. It has taken me a long time to reach this functional unified theory, if that is what you want to call it. For example, in *Dreamstar* I postulated that galaxies reproduce via a space-time galactic cone, which is TOTALLY silly. Thankfully, nobody read that book. With that said, to the best of my knowledge, the universe seems to be a collection of self-replicating battery systems, namely galaxies. Anyway, this degrading energy storage galactic system would explain the randomness of the universe as well as the relative movement of the galaxies within the universe. In short, each subsequent generation of the "Battery System" would be degrading based upon the external environment, and they would also be moving relative to other components of the battery system.

With all that said, it will probably be difficult to see the pattern of this self-replicating galactic battery system because of all the dead branches. By that I mean, fragments of GS1 that: (1) did not contain enough energy to create a black hole recycling bin, (2) didn't fragment to form a spiral galaxy, (3) were inhibited by the external environment, (4) accidently ran into each other, as galaxies tend to do on, (5) Darth Vader ordered the destruction of a particular galactic tree limb with the Death Star, which ironically doesn't destroy stars. As you can pretty well surmise, there could be a lot of reasons why the galactic battery system does NOT undergo galactic self-replication to form smaller galactic battery systems. Of course, I could have called the whole process multiple galactic mitosis, but then I'd have to talk about the difference between mitosis and meiosis, which I'd rather avoid. What I do want to talk about is the possibility of non-galactic non-replicating black holes that are not in spiral galaxies. I guess it is possible that through the complex movement of all the galaxies in the universe, a magnetically complex region could exist that sheds energy and could be labeled as a black hole, but these pseudo-black holes probably don't produced stable galactic battery systems. Actually, these pseudo-black holes are probably only good at destroying energy and adding to the universe's distribution.

In conclusion, there are still quite a few loose ends that need to be tied up with regards to this universal self-replicating battery system postulate. For starters, we still need to deal with the magical-gravity-thingy that aligns planets into solar systems, holds the galaxies together, and creates the black hole recycling bin. Secondly, we live in a Solar System that still remains nameless. Seriously? Hasn't anybody ever thought about naming our Solar System? I mean, what if intelligent life reaches Earth and asks WHAT we call our solar system? We'd be like: "Um, we call it OUR Solar System." To which they'll either pack-up and leave or kill US and take our resources. Actually, that was the most important loose end. I guess I'll address the other loose end in a different essay.

# Essay #7: Gravity

Supposedly, Sir Isaac Newton discovered gravity when an apple tree dropped an apple on his head, which seems a little suspicious. For starters, why did it take a good conk to the head for Newton to realize his placement on this planet? Did he not realize that he wasn't floating about the apple tree? Secondly, Newton was of the mindset that the universe was a CLOSED system, which resulted in a gravitational equation that contained a UNIVERSAL constant? Maybe we should rethink the perspective by which Newton created this gravity equation.

For a long time, I have known that there was something wrong with the theory of gravity and subsequently, the gravitational equation. For starters, the mass of the planets in our solar system does NOT **directly** correlate to the gravity of each planet. Then of course, there is the gravitational equation's total neglect of the concept of temperature. And finally, I have recently postulated that the universe exists as a spectrum of energetic quanta, which means that gravity is **NOT** just ONE energetic quantum but a spectrum of **overlapping** energetic quanta. WAIT, scientists don't think that gravity is an energetic quantum?! Oh, the humanity of an isolated magical energetic force that doesn't make any sense! With that said, in order to explain my postulates about gravity, I have to explain gravity as it pertains to atoms. Granted, you might think this counter-intuitive, but that is the nature of the gravity beast.

The basis of my unified theory is that the universe exists as a spectrum of battery systems, which can be extrapolated to the smallest parts of the universe. To begin with, Einstein postulated that matter is essentially energy. Unfortunately, this has resulted in scientists looking for a **particle** that holds all the energy in matter together. Personally, I think it would be more important to look for the pattern by which the energy within matter is intertwined, but that is just me. Thankfully, I don't have to argue that point to point out the flaws in the current theory of gravity. In short, all you need in order to understand my postulate on gravity is the thermal expansion of matter.

Most of us know that matter expands when it is heated and contracts when it is cooled. Every day we see the cracks in the pavement and feel the potholes, which are the result of thermal expansion and/or contraction. With that said, all you have to do to understand my postulate on gravity is envision WHAT thermal expansion is doing on the atomic level and then cross correlate this event to the directional movement of atoms. Easy, right? So let's start off with the simplest concept, the thermal expansion of atomic orbitals.

Figure 4: Thermal Expansion

When the temperature of an atom is increased, the atom increases in size. And because atoms are made up of a collection of atomic orbitals, the atomic orbitals are the entities that are increasing in size. Now I could get into a theoretical debate as to why, who, when, what, and HOW thermal expansion occurs as it correlates to atomic orbitals and external energy, but I'll save that for latter. The short and skinny is that the thermal expansion occurs. Now all you have to do in order to understand my postulate on gravity is <u>correlate</u> directionally specific atomic orbital expansion to the directional movement of atoms.

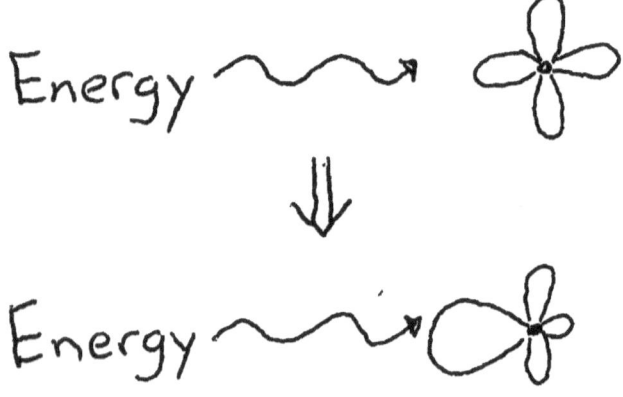

Figure 5: Directional Atomic Orbital Expansion

You might not know this, but most elements prefer to keep their atomic orbitals somewhat **symmetrical**. That means, the FORCE of gravity to MOVE atoms is **NOT** coming from the object producing the "gravitational" energy, but it is the result of the affected atom moving toward the object releasing the "gravitational" energy. It is quite simple if you think about it. The old theory of gravity dictated that large objects are releasing magical-gravitational-energy, which pulls atoms toward the larger object. My theory of gravity dictates that large objects are releasing **energy**, which distorts the symmetry of the atomic orbitals in the affected atom. Then, the affected atom simply MOVES toward the energy source to restore the atom's atomic orbital symmetry.

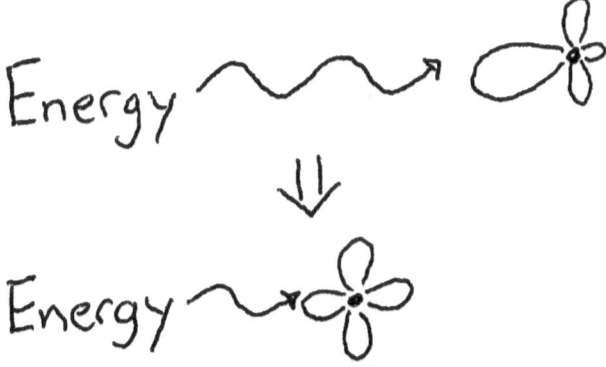

Figure 6: Atomic Movement to Restore
Atomic Orbital Symmetry

With all that said, it is important to note that gravity probably exists as a spectrum of energetic quanta: Thermal and Magnetic. The thermal energetic quanta cause the distortion of an atom's atomic orbitals such that the atom moves to restore atomic orbital symmetry. Granted, the magnetic energetic quanta are probably doing the same thing, but they are also creating a layer of air to prevent the movement of objects away from the thermal source. Quite simply, these two energetic quanta are intertwined to result in the movement of atoms toward the object releasing these energetic quanta and to prevent the movement of atoms away from the object releasing these energetic quanta. Unfortunately, this new postulate about gravity will not be accepted unless I refute Einstein's postulate of a space-time continuum. But as always, I beg to differ.

There is a famous experiment where scientists took two

clocks and **proved** Einstein's theory of relativity, which somehow proved Einstein's postulate of space-time. In short, scientists put one clock on the ground and then flew another clock around in an airplane. When the two clocks were rejoined and compared, they differed by a few nanoseconds or something close to that. With this experimental data in hand, the concept of relativity was thought to forever cement the concept of a space-time continuum. Fortunately, science is more like a glass than cement. By the way, glasses are defined as solid-liquids and cements as solid-solids, just in case you were wondering.

Here is the gist of the two-clock experiment: The only difference between the two clocks, one in the air and the other on the ground, is that each clock was EXPOSED to slightly **different** magnetic field. With this in mind, how about we re-imagine this experiment?

Let's say that we have a clock that only consists of a proton, which is sitting on your living room mantel. Actually, you probably already have a few proton-clocks there, but I don't suggest you go looking for them unless you've got an electron microscope. Next, let's say that the ORDER or ENTROPY of the proton-clock can be used to measure the relative time. Let's say that when you first look at your proton-clock, it is PERFECTLY ordered and arranged. But, after a day of grocery shopping and the like, you come home and the proton-clock is LESS ordered and arranged. Therefore, as you might be able to surmise, you can gauge how much order will be lost in the proton-clock over two days, as long as it remains on your mantel. In short, you have a way of keeping track of OR measuring time as it relates to the environment AROUND the proton-clock on your living room mantel.

*Didactic Essays*

The question that was **NOT** prompted by Einstein's theory and the subsequent clock experiment was this: Is the time variance between the two clocks simply the product of magnetic field variance? I postulate that the components of the gravitational continuum have an intricate effect on the space-time field, which directly affects the period of the proton-clock and subsequently the ability of the proton clock to keep consistent time. Or more simply, a proton-clock sitting on your living room mantel experiences a different magnetic field than the proton-clock in the stratosphere, which will affect the rate by which the two proton-clocks LOSE order and/or registers time.

For example, a proton-clock kept in the stratosphere will experience LESS of the Earth's magnetic field and will subsequently lose less order, which will be read or recorded as the proton-clock experiencing less time. Unfortunately, scientists of yore ERRONEOUSLY associated the difference in **speed** between the two clocks to be the CAUSE of the time variation. All of that resulted in scientists postulating that a person traveling at an enormous velocity will experience LESS time in comparison to a person who is not moving. Of course, it is debatable as to the effect of zero gravity on the human genome, but if a person on a spacecraft experienced the same magnetic field as a person on Earth, then both people will probably age at the same rate.

It is unfortunate that the wonderful theory of relativity in combination with one experimental test, which incorrectly correlates an entropic event to velocity INSTEAD of magnetic field variation, has spawned such a distorted view of the universe. Granted, someone might point out that a theoretical proton-clock is **NOT** the same as the clocks used to '"prove" relativity, but I beg to differ. I've already written a couple books on the repetitive nature of energy in atoms, which seems to be correlated environmental variance. Therefore, I think it would be a little redundant to point out the inner workings of an atomic clock and the similarities to the energy in protons…don't you?

Unfortunately, there is ONE problem with regards to this example: Nobody knows the **exact** magnetic environment that will cause matter to lose order the **slowest**. For example, if the magnetic field in the stratosphere causes a <u>decrease</u> in order lost by the proton clock, then the proton clock will appear to have experienced <u>less</u> time. On the other hand, if the magnetic field in the stratosphere results in an **increase** in order lost by the proton clock, then the proton clock will appear to have experienced **more** time. In short, very little is known about WHAT types of environments and the energetic quanta there in will result in variations to the entropy within protons or matter. All of this is the result of quantum mechanics, math, and how energy is thought to be arranged in matter.

Now that I've somewhat diffused the arcing jeers of my colleagues by recalibrating their perspective with regards to the experiment that "proved" relativity, it is time to think much bigger. "How much bigger?" you might ask? How about our UN-named solar system? That is pretty big, isn't it? How could I correlate this new energetic quanta theory of gravity to our solar system? Think Winnie the Pooh, Think! Oh hey, how about I postulate that the planets in our solar system move about the Sun as a result of the Sun's three-dimensional magnetic flux lines, but the spinning of each planet is the result of energy diffusing from the Sun? Hmmmm, that was an okay thought. Maybe I should expand upon it.

If you were to investigate our universe, you might find out that most of the planets have a magnetosphere. For

*Didactic Essays*

those of you who don't know what a magnetosphere is, my suggestion would be to keep investigating our universe. (Just kidding.) In short, most planets are big bar magnets that create an energetic cage about the planet, which is called the magnetosphere. Of course, it could be a fluke that the planets in our solar system are arranged like iron filings about a bar magnet, but I doubt it.

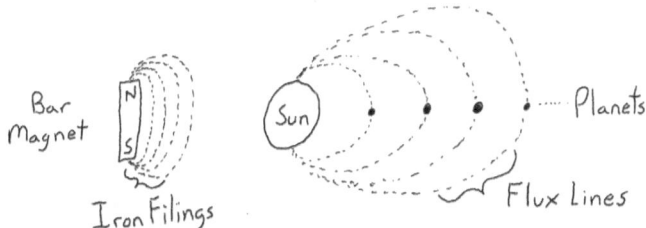

**Figure 7: Planetary Iron Filings**

One reason why our solar system doesn't exactly mimic the nice orderliness of iron filings about a bar magnet is because our solar system keeps getting bombarded with asteroids. If you don't believe me on this, then try making a perfect arrangement of iron filings about a bar magnet, zipping your finger-asteroid across it a couple times, and seeing what it looks like then. Not pretty, is it? Thankfully, our solar system has maintained a relatively ordered concentric arrangement, with the a few asteroid belts here and there, which could be ascribed to planets existing along the Sun's three-dimensional flux lines.

So where does the concentric arrangement of bar-magnet-planets about the Sun's magnetic flux lines leave us with regards to the new theory of gravity? Well, that is a good question. Actually, here is a better question: What is the force that keeps our planet spinning? At first glance, that question is sort of mundane. But when you think about it, it kind of grows on you. What keeps our planet spinning? Is Earth's spinning simply a pre-existing condition? Or, is the energy that is wafting away from the Sun, the energetic-hand that keeps the spin on the Earth-basketball? That was a sort of a strange metaphor, but I'm going to run with it.

Imagine that your finger is a magnetic flux line of the Sun and that it is balancing the Earth-basketball. When the Earth-basketball is spinning, which probably averages out the anomalies in the Earth's bar-magnet, the Earth-basketball is easier to balance. Now, because we live in a place with Entropy, the Earth-basketball will gradually lose its spin, which means there is probably a force that maintains the spinning of the Earth-basketball. I postulate that the Sun's wafting energy-hand gently keeps the Earth-basketball spinning. Of course, this has nothing to do with thermal component of gravity, but who knows how the variable heating of a massive rock will affect its movement. On second thought, let's ponder something closely correlated to the heating of a massive rock, specifically Earth's Magnetosphere.

For a while, some of us have known that Earth has a magnetosphere, which is the result of Earth being a big bar magnet. Unfortunately, that is about the extent of our knowledge. Wait, I think somebody postulated that Earth's core is made of super-dense iron, which creates Earth's big bar magnet, but nobody knows **what** or **why** iron has the propensity to release magnetic energetic quanta. I guess this is where I step in…hello.

One of the first reasons I began wondering about the correlation between gravity and temperature is because the Earth's core is really hot and it yields gravity. Granted, the Sun is hotter and exudes more gravity, but that is beside the point. Actually, there is a point to be made about the

*Didactic Essays*

dissonant nature between the Sun and Earth's core as it relates to energy containment or release, but I'll make that point later. For now, the only thing we need to realize is that the Earth's core is really hot and dense. Unfortunately, this leads to the following question: Is the density/pressure about the Earth's core a factor in the release of magnetic energetic quanta, which creates Earth's magnetosphere? And as always, is there a spectrum of magnetic energetic quanta? As you can see, there are a plethora of factors and very little knowledge as to how these factors are pieced together. One of the reasons why these factors seem so difficult to piece together is because of the first law of thermodynamics, which could be blamed on Newton's idea of a mathematically closed universe.

If magnetic energetic quanta are NOT degradations of matter, then <u>magnetic</u> energetic quanta are SIMPLY **magical**-energetic quanta that can be ascribed a UNIVERSAL constant…or something like that. Therefore, in order to understand the 'battery systems' of the universe, we must be able to calculate in environmental factors. When we are able to calculate in environmental factors to the degrading release of energy from matter, such as the release of magnetic energetic quanta, then things like temperature and pressure become FACTORS in the release of energy by matter. Unfortunately, when we are allowed to factor in environmental factors, the possibility of a spectrum of magnetic energetic quanta comes into light, which makes things a whole lot more complicated.

With all that said, where does all of this leave us with regards to the postulate that gravity is made up of multiple energetic quanta? Well, we know that hot things exude more gravity than cold things. Also, hot things usually have magnetic fields. Therefore, all we have to do is correlate these two ideas to the atomic orbitals in the Earth's core to understand how pressure and temperature are related to the release of magnetic energetic quanta. Seems pretty straight forward, right?

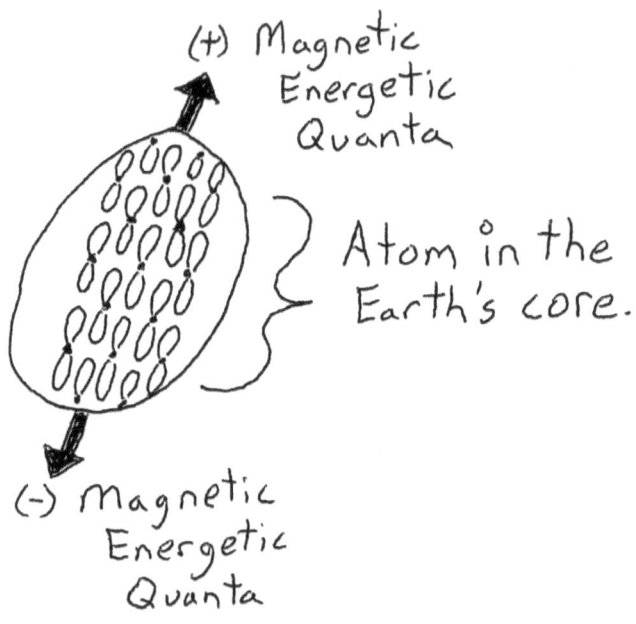

Figure 8: Release of Energy from Directionally Specific Atomic Orbitals

If we know that the Earth's core is really hot and dense, then this means that the atoms are closer together. If the atoms are closer together, then this means the atomic orbitals are closer together. And if the atomic orbitals are closer together, this means there must be greater cooperation between the atomic orbitals. Or put more simply, as a result of TEMPERATURE increasing the size of atomic orbitals and PRESSURE decreasing the distance between atomic orbitals, the atoms in Earth's core are directionally specific. With the understanding of directional specificity about

*Didactic Essays*

the atoms in the Earth's core, which can also be called order, the only remaining road block to understanding Earth's magnetosphere is through the stimulated release of energy by matter. Thankfully, I have another play to help you understand. On second thought, maybe you're not thankful.

(The curtain opens on Malibu Barbie Land)

As Ken's adorable plumber's crack bathes the kitchen cupboards with the radiance of a full moon, Malibu Barbie stands indecisively in front of her full length closet mirror. Her silk lined panties and her vibrant bra with ruffles bounces back and forth as she tries to decide which outfit to wear.

"I can't believe that Malibu Marcy wore that horrible skirt yesterday. It didn't even match her shoes. Oh, I like this." Malibu Barbie says.

Just then, Malibu Barbie's mansion shook.

"Ken! Ken! I think a tree just fell on the house!" Malibu Barbie says.

"What?" Ken says.

"Get up here Ken!" Malibu Barbie hollars.

As Ken climbs the stairs muttering to him-self, Malibu Barbie throws on her robe and runs into the bedroom.

"What is it?" Ken says.

"I heard a huge bang. Didn't you hear it?" Malibu Barbie says.

"I just thought you were throwing stuff again." Ken replies.

"No you idiot, it came from the roof. Go up there and check it out." Malibu Barbie spits.

"Okay, after the plumbing." Ken says.

"Now." Malibu Barbie barks.

"Okay." Ken replies.

"Now!" Malibu Barbie screams.

"Alright." Ken mutters.

"I said NOW!" Malibu Barbie bursts.

"Ok, calm down." Ken says.

As Ken returns to the hallway and reaches for the attic draw string, Malibu Barbie rushes over to her phone. Ken's footsteps and muffled screams echo down from the attic as Malibu Barbie tries to remember her neighbor's telephone number.

"Mary, go to your bedroom window and tell me what you see." Malibu Barbie says.

"Hold on a second. Oh, Barbie, I told you I don't like it when you trick me into looking at your naked body!" Malibu Mary says.

"What do you mean?" Malibu Barbie says.

"Aren't you on the roof sunbathing naked?" Malibu Mary replies.

"No. There is a naked woman on my roof?" Malibu Barbie

"Yeah. Oh there you are. Hi!" Malibu Mary says.

"Hi. I love your bra." Malibu Barbie says.

"You do? I got it from Leman Marcus," Malibu Mary says. "Oh, Ken is on the roof now."

"Hold on, I'm going to three-way with Joe." Malibu Barbie

"Why is there a naked girl on your roof?" Malibu Mary asks.

"Joe, you gotta come over right away. There is a naked girl on my roof!" Malibu Barbie says.

"Ken is trying to peal that naked girl off your roof." Malibu Mary adds into the conversation.

As Malibu Barbie runs down stairs, the doorbell rings. At that precise moment, a massive explosion tears through Malibu Barbie's roof, attic, and into her bedroom.

"Oh my God!" Malibu Mary crys.

"What just happened?" Malibu Barbie asks.

"Ken started humping that naked girl and then they just exploded!" Malibu Mary says.

"Is my closet okay?!" Malibu Barbie hollars.

Malibu Barbie wrenches the front door open, and she and Joe run up the stairs.

"I think so," Malibu Mary says, "but I can see into your bedroom now."

Moments later, Malibu Barbie and Joe are staring out the gaping hole in Malibu Barbie's mansion.

"Maybe that naked girl was...oh my God!" Malibu Mary crys.

Another naked Malibu Stacey falls from the sky and through the hole in Malibu Barbie's mansion. The naked Malibu

Stacey knocks over Malibu Barbie and pins her to the floor with her crotch.

"Oh my goodness! Are you alright honey?" Malibu Joe says.

"Get her fucking va-jay-jay off my face!" Malibu Barbie demands.

For a moment, Malibu Joe ponders the oddness of the situation but then straddles Malibu Barbie and tires to lift the fallen Malibu Stacey off. At that precise moment, a Malibu Stacey arm falls from the sky and wedges itself in Malibu Joe's butt cheeks.

"Oh Barbie! Why didn't you tell me that you were into this?" Malibu Joe moans.

"Get her off of me!" Malibu Barbie screams as she begins to kick, which wedges the Malibu Stacey's arm farther up Malibu Joe's butt.

After Malibu Stacey's arm has completely disappeared up Malibu Joe's butt, Malibu Joe lowers his head to Malibu Stacey's left butt cheek, Malibu Barbie lets out a horrific scream, pushes Malibu Joe off her, tears the fallen Malibu Stacey in half, and gets up.

"If I find out who" Malibu Barbie starts to say, "Joe, you have three arms."

With Malibu Stacey's arm protruding from Malibu Joe's chest, he gets up, takes a pack of cigarettes from his pocket, hands a cigarette to the third arm, lights the cigarette with the other hand, and takes a long drag.

"Don't be so judgmental Barbie." Malibu Joe says as he savors the cigarette.

*Didactic Essays*

"Can I have one?" Malibu Barbie asks.

"Sure honey." Malibu Joe says.

A pair of Malibu Land police officers skids into the bedroom's doorway.

"What the hell!" says one of the police officers.

"Is that you Barbie?" the other police officer asks.

"Yeah, what is so funny Joe?" Malibu Barbie says.

"You've got a goatee honey." Malibu Joe responds casually.

The Malibu Land police officers fall over laughing hysterically. Malibu Barbie whips around and sees her reflection, and then the fires of hell begin to burn in her eyes.

"Noooooooo!!!" Malibu Barbie screams to the sky as fire engulfs her bedroom, her mansion, and Malibu Barbie Land.

(The curtain closes to silence.)

The wonderful thing about just having told a story that explains HOW Malibu Barbie Land can be **stimulated** into a nuclear explosion by two and an eight Malibu Staceys is that it's all relative! For example, if you believe that magnets release magnetic energetic quanta, which exist within a spectrum of magnetic energetic quanta, then you'll understand how Malibu Stacey COULD be a magnetic energetic quantum. All of that would explain HOW magnetically complicated black holes shred energy. Maybe I should explain that in a little more detail…just in case.

My unified theory of the universe states that there is a spectrum of battery systems. For example, black holes are really good at shredding energy, which means there must be ENERGY that is **stimulating** the decay of matter within black holes. And because we know that most planets and stars are really big bar magnets that are releasing magnetic energetic quanta, then it is possible that someone might correlate the **stimulated** degradation of matter in black holes to the extremely complex arrangement of magnetic energetic quanta about the black hole region. Wait, did I just postulate that? Anyway, if magnetic energetic quanta from stars are able to **stimulate** the degradation of matter about the black hole region, then this is also method by which energy can be RE-organized to produce Galactic Seeds. Thus, a galaxy is nothing more than an extremely complex battery system.

Well, that about sums this essay up. Hopefully, you have learned that gravity is NOT a magical-energy that radiates from matter <u>regardless</u> of the temperature and/or pressure. More than likely, gravity is combination of thermal and magnetic energetic quanta spectrums, which cause the distortion of an atom's atomic orbitals such that an atom **<u>moves</u>** to restore atomic orbital symmetry. Anyway, the use of a UNIVERSAL constant might be a good approximation of gravity, but it completely overlooks stimulated release of energy by matter, which is dependent on the environment.

# Essay #8: Hysterical Electrons

I get the feeling that people think black holes are evil. Of course they are mysterious, but that is because they don't emit the types of energy we are used to seeing, which is electromagnetic radiation for those of you who were wondering. In fact, if you think about it relativistically, black holes are VERY similar to matter. For example, both contain energy, are dark unless stimulated, and eventually degrade. There, do you feel somewhat better?

Even though I've already pointed out that the periodic table mimics a spectrum when all the isotopes are added to it, it is time to delve deeper into the possibility that ALL matter exists as a spectrum. Granted, we might not be able to measure this variance, but matter can release ENERGETIC photons. I mean, didn't Einstein come up with an equation that correlated energy to mass, which means that matter is losing mass by emitting energy? Hmmm, let's ponder this new and exciting possibility of a spectral universe.

If one were to look in a science book these days, you'll probably find the exact weight of protons, neutrons, and electrons, which will seem REALLY accurate because these measurements are recorded to the umpteenth place. But, when you realize that ANYTHING that has ENERGY must have **MASS**, you may begin to wonder. For example, you might wonder if all electrons are the same, even though some electrons are EXCITED. If you have pondered that, then here is yet another play to enlighten your quandaries.

(The curtain opens to hysterical environment.)

"Imagine a TV show, kind of like Charlie Rose meets Jerry Springer, where the host interviews atoms and their electrons," says the announcer.

"So tell me Mr. Carbon, why do you have so many electrons?" the host asks.

"The little bitches just dig me," Mr. Carbon says.

"Well, have we got a surprise for you. All your electrons are here today!" the host says.

"Ohhhhh, damn! I didn't sign up for this shit!" Mr. Carbon blurts.

"Shut your fucking trap you dumb shit! I've been hanging around you for over four million years and you've NEVER returned any of my phone calls," 1S-electron says.

"Oh snap girlfriend. He's been my man since the beginning of time!" Another 1S-electron says.

"Fuck you, you dirty whores! I saw his charges first and I'm the only one that loves him in spite of his addiction to you dirty charge whores," 2S-electron syas.

"Ladies, I love you all the same!" Mr. Carbon replies.

"Oh shut the fuck up. I know you spend more time with that tramp and her lose 1S-electron charge," 2P-electron spouts.

"Ladies, thank you for joining us today. Please sit down.

*Didactic Essays*

There will be plenty of time to discuss Mr. Carbon's short comings," the host pleads.

"Who do you think you are with that smug little smile? You'll never understand how to treat an electron and the way he's excited me!" 2P-electron shouts.

"Excuse me, but I'm hosting this TV show," the host says.

"Well you can shut the fuck up and get out of my way. You're blocking my view of my sexy man charge," 2P-electron says.

"Mr. Carbon, do you love all these electrons the same?" Host asks.

"Well you know..." Mr. Carbon replies, but gets cut off.

"Oh you better not say you love those 1S-electron whores!" 2P-electron shouts.

"I'm Mormon, what can I say?" Mr. Carbon says.

"Oh hell NO! You said you were Catholic!" 2S-electron spits.

"I was baby, but I converted." Mr. Carbon replies with an electronic shrug.

"He converted for me because he loves me the most!" 2P-electron says.

"Shut the fuck up you dirty tramp!" 1S-electron

(Flying telephones, chairs, and a rubber ducky that says: "Good day mates. I'm a zebra rubber ducky from down under, and I've got no clue as to how I got involved in this war or how electrons can muster up the energy to throw things. When I was a little rubber ducky, my mates told me about such silly TV shows like this, but I never believed them. Cheers!")

"Ladies please! Let Mr. Carbon answer the question," says the host.

"Damn bitches!" Mr. Carbon says as he rubs his head.

"Please Mr. Carbon, answer the question." The host pleads.

"You fucking hit me in the head, bitch!" Mr. Carbon says. "Ok, give me a second. I love them all the same, but in a different way."

"Asshole, I'm finding myself a new atom. And for the record, I could barely see your tiny charge!" 2P-electron says.

(The audience claps, cheers, and becomes ligands?)

"No, no. I really love each of these sexy electrons, but for different reasons. I love the 1S-electrons because they are always around and soft. I love the 2S-electrons because they are a little more firm with my charge movement. I love the 2P-electrons because they are tough as nails and they're only good for a fling every once-n-awhile," Mr. Carbon says.

"I don't fucking love you, you charge-whore! I was just hanging out with you because you were a rebound atom," the 2P-electrons shouts.

"So what you're saying Mr. Carbon, is that these sexy electrons have varying levels of hardness that correlates to their intrinsic energy and that you enjoy each of them in different situations?" the host asks.

"Yeah" Mr. Carbon answers.

"Well, that is all the time we have today. Tune in next

week when we investigate inter-ionic relationships," the host concludes.

"If you know any electrons that have been part of a covalent bond or a double bond and wish to be on this program, please mail us your story at the following address," the announcer says.

(The curtain closes on this absurdity.)

With this explicit play in mind, you might be wondering a couple things. For starters, how does an electron increase in energy such that it becomes EXCITED without a change in **mass** or ENERGY? Wouldn't that go against Einstein's equation? Next, how does an electron release a photon, which has energy and mass, without losing any ENERGY or **mass**? Unfortunately, the only way to address these quandaries is to imagine that EVERY electron is exactly the same.

For the sake of absolute simplicity, let's imagine that EVERY electron is EXACTLY the same. Now we could call every electron an Average Joe, but the **word** average means that there is a range of variance about the electrons. Maybe we should just call **every** electron in the universe, Joe? Hey look, there's Joe in the 1S atomic orbital and Joe is in the 2P atomic orbital too. Damn it, this place is teaming with Joes...not cool. Point taken?

For the sake of absolute complexity, let's hypothesize that there is a spectrum of electrons. If there is a spectrum of electrons, this will only lead to more complex and annoying questions like the following:

1. If there is a spectrum of electrons, can only

certain electrons occupy certain atomic orbitals?

2. When an electron releases a photon, does the electron change in **mass** or ENERGY?

3. Do photons magically convert to mass when they hit electrons?

4. If photons are "massless," then how do they affect electrons, which have relatively enormous masses?

5. Does the interaction of a photon with an electron depend on the energy of the electron, the electromagnetic state of the atomic orbital, or the type of photon?

As you can see, the main reason why nobody has postulated that protons, neutrons, or electrons exist as a spectrum of energetic entities is because nobody wants to deal with the ensuing questions, myself included. Actually, I'll deal with those questions in another book.

Now that you're slightly irritated and fuzzy to cognition, I think it is a perfect time for YOU to ask me some questions like these: What happens to electrons when they kick the bucket for the last time? Does the destruction of unique electrons create unique amounts of plasma? Do plasma quanta keep degrading into other energetic quanta? And finally, how does all of this intertwined into the unified theory of a universe made of battery systems?

In all honesty, I'm glad that YOU asked me those questions because I'm running out of transitions. Within my unified theory of a universe made of battery systems, each layer of

*Didactic Essays*

the Energetic Pyramid acts as an insulator. For example, the presence of plasma quanta prolongs the degradation of thermal energetic quanta. Or the presence of thermal energetic quanta prolongs the degradation of electromagnetic energetic quanta. Or the presence of electromagnetic energetic quanta prolongs the degradation of magnetic energetic quanta. Or the presence of magnetic energetic quanta prolongs the degradation of matter. Or the presence of matter prolongs the degradation of the energy in stars. Or the presence of stars prolongs the degradation of the energy in black holes, which subsequently diffuses and replicates the whole system into smaller battery systems.

One unfortunate caveat to the hypothesis of prolonged energy degradation with regards to the hierarchy of the Energetic Pyramid is: **Relatively**. By this I mean that EXTREME amounts of magnetic energetic quanta might not prolong the degradation of matter but actually cause the degradation of matter. For example, the way I explained how the magnetic energetic quanta of stars cause matter to degrade in black holes. And on top of that relativity, there is also the possibility that certain types of energy **prefer** CERTAIN types of energy, which will prolong their degradations.

In conclusion, once scientists have figured out a way to more accurately weight minute quantum of the universe, they will probably find out that ALL protons, neutrons, and electrons are **NOT** the same. Fortunately for scientists, this will only be the beginning of a wondrous realm of exciting questions, all of which, will be theoretically <u>massless</u> but contain LOTS of **energy**?

# Essay #9: Quanta Dynamics

Thus far, I have used the words 'quantum mechanics' to label my quantum theories. Unfortunately, quantum mechanics is defined as the movement of quantum particles as they are understood by MATH...MATH! (I said the second MATH like Seinfeld always said Newman, just to annoy all the scientists.) Fortunately, because I've already pointed out the theoretical void that MATH creates within science because MATH is just the result of the repetitive patterns within the universe, it is time that I use a different term. From hence forth, Quanta Dynamics will be the study of quanta particles in a dynamic universe...even if it is just me thinking about it.

In addition to Quanta Dynamics being about the study of quantum particles existing in a dynamic universe, it will also be about how energy can degrade. For example, if you walk outside and bask in the light of a star, you will feel warmth. Now the old theory of light induced thermal elevation dictates that the increase in temperature is the result of light increasing the speed of the electrons in your skin. Unfortunately, the concept of Hysterical Electrons does not **FIT** Einstein's famous equation, $E=mc2$, but it does fit into Quanta Dynamics. You see, within Quanta Dynamics, quantum particles are able to **degrade**. Therefore, under Quanta Dynamics, when you walk outside and feel

*Didactic Essays*

the warmth of the Sun, it is the result of electromagnetic energetic quanta degrading into thermal energetic quanta. Actually, there are tons of examples of energy degradation in science.

For a while, scientists have been able to detect these quantum particles called Neutrinos, which are energetic degradations of fusion. Granted, the theory behind the detection of Neutrinos is a little bit outdated, but that is just because they are based upon Newtonian physics. You see, Newtonian physics is based upon the concept that all energy in the universe acts like billiard balls, which is totally boring. I mean, if energy were non-degrading billiard balls, then there would not have been a nuclear arms race…or something like that. Anyway, because a picture is worth a variable amount of words, here is my theory on Neutrino detection, which is within the realm of Quanta Dynamics:

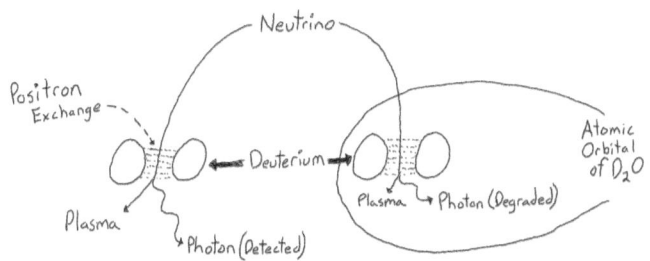

**Figure 9: Neutrino Degradation**

From this figure, you might not be able to extract the point that the Deuterium atom is in the atomic orbital of $D_2O$, but that is why I said pictures are worth a variable amount of words. With that said, as a result of the Deuterium atom being in the atomic orbital of $D_2O$, the Deuterium atom can undergo what scientists call "Hydrogen Bonding," which is kind of confusing since the Deuterium atoms have replaced all the Hydrogen atoms in $D_2O$. Regardless of that semantic point, Deuterium bonding occurs when the Deuterium atom changes atomic partners. By that, I mean that the Deuterium atom is leaving its current atomic orbital and entering the atomic orbital of the adjacent atom. One consequence of Deuterium bonding is that for a brief moment, the Deuterium atom is NOT surrounded by an electron cloud. I postulate that in this brief moment of FREEDOM, Deuterium is free to catalyze the degradation of a Neutrino into a photon WITHOUT the electron cloud causing further degradation to the photon. In short, when the Deuterium atom is within an atomic cloud and catalyzes the degradation of a Neutrino, the electron cloud causes the further degradation of the photon, which renders it undetectable. All of which is extremely wordy and boring. How about a mental song break?

"OH-ahhhh unique POSITRON environment…crap." The Neutrino Song.

In conclusion, it is amazing how variable the universe could be if we look at it Qunata Dynamically. Granted, the idea that the universe exists as a spectrum of energetic quanta, which are continually degrading, is going to complicate the way we SEE the universe. Butt, I'll save all those words for another book. Anyway, I hope you'll give the Ivory Tower and science a second chance, if you find the relative time. The universe always looks a little bit different when you LEARN something and then look at it again. Who would have ever thought, right?

# Essay #10: The Sparkle of Darkness

No matter how I cover my eyes, the sparkle of darkness remains. I can hear the world around me as my photo-receptors tingle with anticipation. They are the most sensitive receptors known to man and beast.

As my eyelids crack, a blurry world comes through my lenses, which were fashioned over thirty years ago. I know that the light is refracted depending on how much water covers my eyes. I know the light is refracted depending on what types of proteins are in my eyes. I know my perception of detail is dependent on the proximity of the photo-receptors within my eyes. The world comes into focus and details are split, processed, combined, and identified within my head.

I see a light glisten and glitter behind a tree. The leaves are dark, the trunk almost black, and the world is its halo. I perceive the world and the funny things jumping around it. I know I must seek food. I know I must seek water. I know I must seek shelter. But right now, I delay my chemotaxis to enjoy the radiance of light about one lonely tree.

As I shut my eyes, the glitter of what was there, haunts my eyelids. The world hasn't changed with the blink of an eye, but my perception has. I know I must open them again, but I am afraid of what I might find: Negligence, stupidity, or indifference. The fear of our reflective existence echoes about my neurological synapses. If the sight of pain doesn't fill me with emotion, then I have successfully filtered the world to fit the picture of how I want it to be. No amount of tears will ever change this.

# Essay #11: The Tower Perspective

1) There is a magnificent tower. It is made of pure ivory and reaches past the clouds. The people inside the tower are snobby, self-centered, and extremely odd.

2) There is a building under construction and it is falling down. It is made of people, books, and dreams. The people outside the tower are religious, angry, and often irrational.

3) There is this imaginary place, a place where extremely odd people explain books to the religious, angry, and drunken young people. It is a place where individuals are struggling to define, redefine, identify, ponder, muse, learn, help, and think.

4) There is this chair that is uncomfortable, a chair that is hard, lumpy, and falling apart. When people sit in it, they can't get comfortable. They squirm back and forth, trying to understand why they are so uncomfortable. Only the chronically uncomfortable find the chair comfortable.

5) There is a car that doesn't run very well. The engineering area is okay; neglect has covered it in rust, and children cower in the back seat, afraid to be seen. The owner is old and wants nothing to do with those fancy new electric cars.

6) There is a place in all of us. It is a place we build to contain everything we think we are. It is often a place of disuse, broken dreams, and broken chairs. It is a place of Entropy. Some call it Ego.

7) There is an Ego. It is a place we build to feel happy about how we fit, interact, and exist in the world. Truth deflates the Ego and Lies inflate the Ego.

8) There is a planet. It is a place of thought, reflection, trust, distrust, anger, hunger, fear, laughter, Ivory Towers, broken down shacks, old cars, uncomfortable chairs, the religious, the agnostic, and OUSISTs, which is a mix of everything and nothing at the same time.

# Essay #12: Do NOT Tilt!

No matter what universal theory you adhere to, there is no question about the presence of gravity and entropy. Unfortunately, we have wasted a lot of time debating HOW the universe has reached this mixture, but that isn't really important. What is important for the Earth SPECIES is the behavior of the universe as it is HERE and NOW. Thus, it is important that we keep looking to the starry night sky. Butt, what determines what we see?

Quite simply, we know that light doesn't travel through cold matter. We know that cold matter doesn't emit appreciable amounts of energy. We know that stars can bend light. We know that heat and/or hot gas can distort light. We know that black holes display enormous amounts of energy/gravity and can distort light. We know that magnetic fields are created by magnetic moments, which are energetic quanta that interact with light. And finally, we know enough to know that we do NOT know everything.

All of this means that, we should be asking the following question: Do we live in a MOVING pinball universe where light is NOT always coming straight from its source?

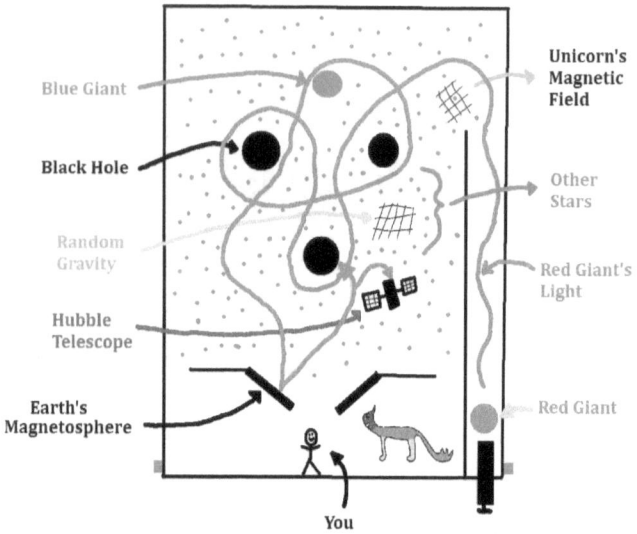

Figure 10: Moving Pinball Universe

If we existed in a moving pinball universe where everything is simply a distortion of a distortion and we have just stumbled upon this knowledge, then we need to stay CALM. Remember, we still have each other. WAIT. That's not very comforting. How about we remember that we have the power to change things for the BETTER! Wait, that means we also have the power to change things for the WORSE. I guess it all depends on HOW the things we see are manipulated or HOW the words we use can be manipulated. With that said, the pinball of knowledge is in motion and unless you add to it, it will DEGRADE like all the other energetic quanta in this universe. Ooops, the cat is out of the bag?

# New Age Adages and Addendums

TV commercial: In a world where words mean more than actions, actions mean more than turds, and turds are always talking about actions, one lone essay dares question the presumptive conceptualization of time tested adages. There are romantic word meanings; There is horrific conceptualizations; And there are the roots of shit that doesn't matter anymore. Tune into this epic premiere…in a couple spaces.

1. Mind over matter. - This adage is delightfully close to being correct. Cognition based neurotransmitter release and variable genomic expression can only be accomplished through electron modulation within proteins, amino acids, and atoms. Unfortunately, this adage doesn't note the limitations of mind over matter. Therefore, here is a more accurate adage to live by: "Mind over matter except when you're vitamin deficient, starving, dehydrated, ignorant, genetically compromised, or dead."

2. Sticks and stones may break my bones, but words will never hurt me. - This adage is repeated by millions of mothers and fathers,

but nobody seems concerned that it isn't really true. Here is a more accurate adage to live by: "Sticks and stones may break my bones, but verbal discourse will forever taint my perception of the universe?"

3. My only regret is that I only have one life to give my country. - This adage assumes that countries are equivalent to God, which is kind of disturbing if you think of it. Here is a more accurate adage to live by: "My only regret is that I only have one life to give my country <u>if and only if</u> my country professes the correct religion, interpretation of that religion, and is winning the fight for world domination."

4. Two wrongs don't make a right. - This adage is supposed to influence people NOT to do wrong, but in the case that ONE wrong is HORRIBLY wrong and you've got to do a little wrong to stop a HORRIBLE wrong, I think two wrongs might make a right. Here is a more accurate adage to live by: "Two wrongs DO make a right when you have the written approval of the Pope, the President, and you're mother."

5. Honesty is the best policy. - This adage was probably fashioned in the time of Abraham Lincoln, but unfortunately the US government doesn't follow this adage anymore. Here is a more accurate adage to live by: "Honesty is the best policy except when it's NOT."

6. A bird in the hand is worth two birds in the bush. - This adage is completely confusing

*Didactic Essays*

in the age of fast food, computers, and Vice Presidents shooting their hunting partners. Here is a more accurate adage to live by: "One burger in the hand is worth two large fries behind the counter."

7. Treat others the way you want them to treat you. - This adage is moderately important to most Homo sapiens, except when people have insane world views. In either case, here is a more accurate adage to live by: "Most Homo sapiens are self-centered, illogical, and very temperamental. Avoid them at all costs because no matter how you treat them, they will always find a way to trample you."

8. Always say thank you. - This is an interesting adage in that you should always say thank you. No matter what? Thank you, America.

9. Turn the other cheek. - This adage is actually too pornographic for conservative America because people think of butt-cheeks when they hear the word "cheek." Here is a more accurate adage to live by: "If you're being spanked, make sure and say: 'Harder baby, harder!'"

10. An eye for an eye. - This adage refers to something other than turning the other cheek, which is mildly confusing because the people who use this adage are often caught using the other adage as well. Here is a more accurate adage to live by: "Turn the other cheek unless they accidentally hit your eye. In which case, poke their eye out."

11. All men are created equal. - Everybody likes to use this adage, but nobody ever believes it. Therefore, here is a more accurate adage to live by: "All men are created equal, but if you didn't go to a good school, then you're shit."

12. Ask not, want not. - This adage is supposed to help those people who didn't get into a good school. Here is a more accurate adage to live by: "Suck it up."

13. The world isn't perfect. - This adage is used by most Homo sapiens as a reason not to do anything. Here is a more accurate adage to live by: "The world isn't perfect, but that isn't a logical excuse for letting it go to hell."

14. A mind is a terrible thing to waste. - This adage is totally laughable in America. Here is a more accurate adage to live by: "A mind is a terrible thing to waste, unless you just need someone to clean-up after your shit."

15. Guns don't kill people, people kill people. - This adage is out of date. Therefore, here is a more accurate adage to live by: "Lasers don't kill people, laser holes kill people."

16. If you're going through hell, keep going. - This adage doesn't make any sense. Here is a better adage: "If you're going through hell, stop and find out what their health insurance covers so you can tell your friends. "You think you've got it bad? Well in Hell, their copay is actually an arm and a leg!""

*Didactic Essays*

17. If you don't have anything good to say, then don't say anything at all. - This adage actually doesn't work in a capitalistic system. Here is a better adage to live by: "If you don't have anything good to say, don't become an actor, anchor, or TV personality."

18. What goes around comes around. - This adage is about Karma and usually inspires people to let bad people do bad things. Here is a better adage to live by: "I'm not fucking with that retard, they're crazy!"

19. Good things come to those who wait. - This adage goes with the adage about not fucking with crazy people.

20. Birds of a feather flock together. - This adage is actually quite true when it comes to birds, but not so much for mammals. Here is a better adage to live by: "Mammals with the same hair arrangement NEVER flock together, except in high school."

21. Opposites attract. - This adage seems to hold true in the realm of quanta dynamics, but doesn't seem to make much sense when looking at a spectrum of Homo sapiens interactions. Here is a better adage to live by: "Bald monkeys don't like bald monkeys, unless their daddy was bald too?"

22. It takes all sorts to make a world. - If Christians had their way, this adage would have been erased from history's ledger...along with the Muslims.

23. Muslim people don't have morals. - This adage is relatively new, but completely wrong. Here is a better adage to live by: "Christian people who believe that Muslim people don't have morals, don't have morals...and so on and so forth."

24. Everything I needed to know, I learned by fifth grade. - Whoever started this adage, obviously didn't attend sixth grade. Here is a better version of this adage: "I learned how to wipe my own ass by the fifth grade."

25. A penny for your thoughts. - This isn't actually an adage, just a quote from a thought-prostitute.

26. Never eat yellow snow. - This adage only holds true unless the snow is from a street vendor. In either case, you still probably shouldn't eat it.

27. Never mix beer and wine. - I never understood this one. Here is a better version of the adage: "If you're drinking beer, nobody can call you a whino."

28. Liquor before beer, you're in the clear. Beer before liquor, never sicker. - This adage was probably started by puritans. Here is a better adage to live by: "Teach your kids how to drink responsibly. Otherwise you'll regret it when they get to college."

29. Rome wasn't built in a day. - This adage is wrong-ish. If God built Rome, it would have only taken a couple minutes.

30. Where there's a will, there's a way. - Here is a

*Didactic Essays*

better adage: "Where there's a will, there's a very loving, caring, and extremely supportive family."

31. Absence makes the heart grow fonder. – Here is a better adage: "Absinthe makes the heart grow fodder."

32. There is no accounting for taste. - This isn't true. The account is actually in Switzerland and I'm sure there are plenty of accountants there.

33. Actions speak louder than words. - Only if you're psychic.

34. After a storm, comes the calm. - Here is a better adage: "After the storm comes the calm, unless you're in the eye of a hurricane. In which case, you're shit out of luck."

35. An apple a day keeps the doctor away. - This adage was deemed UNcapitalistic and was replaced with the following adage: "A happy meal is cheaper than most apples."

36. Ask a silly question and you get a silly answer. - This one isn't true at all. Nobody as ever asked me any silly questions.

37. Attack is the best form of defense. - This one is true unless you can't afford to attack because of your domestic policies.

38. A bad excuse is better than none. - If this was true, then why do parents still rely on the 'because' excuse?

39. A bad workman blames his tools. – And what? A good workman blames God?

40. A barking dog never bites. - I don't think this adage has any research to support it.

41. If you can't beat them, join them. - This adage is actually implying the following: "If you can't beat them, join them and beat them when they're least expecting it."

42. Where bees are, there is honey. - I learned this one wasn't true when I stepped on a bee in kindergarten.

43. Beggars can't be choosers. - I think this adage is actually an adage overlap with: "You can't do two things at once."

44. The best of friends must part. - Nothing wrong with this adage.

45. The best things come in small packages. - Obviously, the redefinition of "package" made this adage obsolete.

46. The best things in life are free. - This adage is completely retarded. It should be changed to this: "The best things in life are free when you can't afford anything else."

47. Better late than never. - This one doesn't apply in any medical field.

48. Better to have loved and lost, than never to have loved at all. - A better adage would be as follows: "It is better to redefine the concept of

*Didactic Essays*

love than have lived your life regretting a stupid definition."

49. The bigger they are, the harder they fall. - In America, we know this one isn't true. We grow them so BIG that they can't get up to fall down.

50. You cannot get blood from a stone. - This adage was the result of alchemy.

51. You can't judge a book by its cover. - This adage is often confused with this one: "People are like open books."

52. If it ain't broke, don't fix it. - This adage only works when you don't have a full report on what will happen when something breaks.

53. Business is war. - This one is completely true.

54. Be careful what you pray for, because you might get it. – Apparently, nobody is praying for peace.

55. A chain is only as strong as its weakest link. - This adage isn't true because if you double up the chain, the chain becomes stronger than its weakest link!

56. Don't change horses in mid-stream. - Unless the first horse drowns.

57. Children of fools tell the truth. - This one is totally true, except in the days of Abraham Lincoln.

58. Civility costs nothing. - This doesn't hold true when business is war.

59. A man is known by the company he keeps. - This is only true for Microsoft.

60. Why buy the cow when you can get the milk for free? – Here is a better adage: "Never let the cows out to give their milk away for FREE."

61. Crime doesn't pay. - This adage needs an amendment: "Crime doesn't pay unless it is corporate crime."

62. Crosses are ladders that lead to heaven. - If this adage is true, then Homo sapiens stopped sending people to heaven around the fourth century.

63. There is no use in crying over spilt milk. - Just get a cat.

64. Curiosity killed the cat. - Actually, I think it was the high levels of low density lipoproteins.

65. The customer is always right. - Until you reach the employee break room.

66. Let the dead bury the dead. - This has been quoted by many Scientologists as proof that Jesus was rebuking the scientific thought of the time, namely that dead bodies stink.

67. Dead men tell no tales. – Apparently, this adage was forged before TV shows about special crimes units.

68. The devil can quote scripture for his own ends. - I've never heard this adage before.

69. The devil is in the details. - Of science?

*Didactic Essays*

70. You can only die once. - Unless someone is there to revive you.

71. Do right and fear no man. - This is backwards: "Do right and fear ALL men."

72. A dog does not eat dog. - This might be true, but I have seen birds eat chicken.

73. Early to bed and early to rise, makes a man healthy, wealthy, and wise. – Not to mention, single.

74. When elephants fight, it is the grass that suffers. - Republicans hate fighting.

75. The enemy of my enemy is my friend. - This adage is only true in politics, war, and sex.

76. To err is human, but to forgive is divine. - This adage needs a spit shine: "To fuck up is human, to forgive is divine, and to trust a forgiven fuck-up will result in a holy shit storm of regret."

77. Every man is the architect of his own fortune. - Unless he never went to architect school.

78. Experience is the best teacher. - Except when the experience is death.

79. Faith will move mountains. - That and coal mining.

80. All's fair in love and war. - Please see adage #48.

81. The fat man knoweth not what the lean man thinketh. - Except if the lean man is thinking about food.

82. Give a man a fish and he'll eat for a day. Teach a man to fish and he'll eat for a lifetime. – Here is the updated version of that adage: "Give a man a fish and he'll eat for a day. Teach a man to charge for fishing poles, fishing licenses, and access to the water, he'll become a capitalist for life?"

83. The only good Indian is a dead Indian. - Wow, I sense a lot of anger in this adage.

84. Hell hath no fury like a woman scorned. – Or a women facing a lifted toilet seat.

85. An idle brain is the Devil's workshop. - Or laboratory?

86. The last drop makes the cup run over. - Unless the thermal energy of the drop causes viscous contraction of the solution.

87. One does not wash one's dirty linen in public. - Unless you're poor and you have to use the local washateria.

88. A watched pot never boils. - Eyes can absorb an endless supply of thermal energy?

89. Waste not, want not. - Except in a capitalistic society.

90. United, we stand. Divided, we fall. - This adage doesn't support a democracy. It should be this: "United, we stand. Divided, we have a democracy."

91. It takes a whole village to bring up a child. - Or an idiot.

*Didactic Essays*

92. Trust in God but tie your camel. - Thank God we don't use camels anymore.

93. Don't throw the baby out with the bathwater. - But it is okay to drain the water to get the kids out of the tub.

94. Don't teach your grandmother to suck eggs. - Unless you're trying to make Easter decorations and she has Alzheimers.

95. Never speak ill of the dead. - Except for Hitler and anyone who didn't die on your time scale.

96. Self-preservation is the first law of nature. - Except when death is unavoidable.

97. Seeing is believing. - Except when you don't believe your eyes.

98. When in Rome, do as the romans do. - Build a coliseum and make criminals battle for their lives.

99. You can't change people. - This one is oh so true. Here is a better adage: "If you can't change people, just ignore them. They are bound to show their true selves at some point."

100. One picture is worth a thousand words. - Except, Microsoft Word's word counter doesn't add in an extra thousand words for each picture.

101. Brevity is the soul of wit. - Unfortunately, too many things have souls and not enough things have intelligence.

102. I thunk bunger, therefore I am a derivative of mindless consumption?

103. Blood is thicker than water. - This adage is supposed to inspire a bond between family members, which is somehow correlated by the obvious viscosity difference between these two liquids. Ironically, if someone is in need of a blood transfusion, it DOESN'T usually come from a blood relative. With all that pensive genealogical idealism, it is amazing that the blood that might someday save your life will come from someone that has a COMPLETELY different phenotypically appearance. Here is a better adage: "Blood is more viscous than water and this is important because?"

104. If at first you don't succeed, try, try again. - This adage is supposed to inspire people to never give-up. Unfortunately, this isn't true in all circumstances. Here is a better adage to live by: "If at first you don't succeed, just assume that you've sinned, God hates you, or it is a governmental conspiracy."

105. For every action, there is an equal and opposite reaction. - This adage was forged when everyone thought the universe was a CLOSED system. Here is a better adage: "The edge of the universe is a reaction free zone."

106. If you love someone, set them free. If they come back, they're yours. If they don't, they never were. - Honestly, I don't know why you have to "arrest" the person you love, but here is a better

*Didactic Essays*

adage: "If you love someone, set them free. If they don't come back, try being less of a control FREAK next time."

107. If Karma is a bitch, then Logic must be a philanderer.

108. You can't please everybody, but you can bomb whomever you please?

109. You can't educate everyone, but you can incarcerate whoever falls through the educational cracks?

110. One must understand the limits of your wisdom before you can comprehend the system in which you exist.

111. If there is NO dignity in being naked, then dignity is strictly dependent on your looks.

112. It is not insane to create atmospheric vibrations when you are practicing a speech, singing, or communicating to other tympanic membranes, but mental cognition through atmospheric vibrations to your own tympanic membranes is absolutely unacceptable!

113. When people say they are brain storming, why don't their thoughts include all the gloom and tears like that of a real storm?

114. A bone to pick with you. – This adage used to be used as a phrase of friendship because the people were sharing the bone at which they were picking…kind of like a peace pipe. But

today, it is a precursor to hostility...kind of weird how phrases evolve.

115. Did the egg or chicken come first? – This has been an age old brain teaser, but with the knowledge that ossification occurred in the earliest of sea dwellers, the logic is that an exoskeleton structure to protect developing organism occurred first, but was later exploited by those crazy land dwellers.

116. Beauty is in the eye of the beholder. – Here is a better adage: "Beauty is in the eye of the largest media conglomerate, which is the result of visual conditioning."

117. Only pea-brains visualize peas?

# Glossary

1. Thought bungers: thoughts with cobwebs on them?

2. Beyond reason: science.

3. Dark Matter: scary stuff that exists in every part of the universe except OUR solar system?

4. Benjamin Button Theory: the idea that the universe will hit the brakes and put it in reverse....right?

5. Entropy: stuff of the second law of thermodynamics.

6. Enthalpy: temperature, which is only stored in variable electron movement?

7. Intellectual intermission:

8. Hysterical electrons: electrons that just found out their Atom has been seeing other electrons.

9. Average Joeverse: a universe where all electrons are ABSOLUTELY the same.

10. Malibu Barbie Land: a weird-weird place with NO order?

11. Twinkie: something my ex-flatmate called me after I moved out.

12. Imaginary thought-drugs: any IDEA marketed by one individual to another.

13. Thought-drugs: any IDEA that makes you feel something or anything.

14. Meandering thesi: my way of confusing you into comprehension?

15. China's heaven: America?

16. Science: the opposite of Religion?

17. Monkey Butt: reader.

18. Capitalistic education: a money making business?

19. Photons: little packets of energy with MASS, but don't tell anybody else.

20. Gravity: something that already has a UNIVERSAL constant?

21. Theories: just ideas.

22. Postulates: the preamble to theories?

23. LOW-GIC: anything other than logic.

24. First law of thermodynamics: WRRRRRONG!

25. Second law of thermodynamics: Right?

26. New Scientific Foundation: something with LESS magical pathways?

27. Mental exercise: any thought that makes your head hurt.

*Didactic Essays*

28. Scientists: slaves to Capitalistic Education or Religion?

29. Closed universe: a place where energy never disappears, but people do.

30. Matter: protons, neutrons, and electrons.

31. Electromagnetic spectrum: NOT matter?

32. Militarized Scientific-Gestapo: publishers of LOW-GIC?

33. Phusion: a cousin of fusion that requires energy to make heavier elements.

34. Fission: energy released by one atom splitting into two atoms.

35. Anti-gravity: anything that is not gravity?

36. Adages: the defendors of LOW-GIC?

37. Exorcist: an individual that excises personal demons with exercise?

38. Warlock: someone with the key to war?

39. Warship: a religious service were worshipers pray for war?

40. Prayerlock: someone whose prayers mean more?

41. Tearrist: someone who cries over spilt milk?

42. Books: mental wormholes?

43. Adages: COMMON sayings or phrases that always need referencing?

# REFERENCES

1. The Oxford American Desk Dictionary and Thesaurus 2nd Edition. New York. The Penguin Group, 2001

2. Jennifer Speake and John Simpson. The Oxford Dictionary of Proverbs 5th Edition. Oxford University, 2008.

3. Ira K. Wolf and Sharon Weiner Green. Barron's GRE: Graduate Record Examination 17th edition. Hauppauge, NY: Barron's Educational Series, 2007.

4. Willie Nelson. On the Road Again. Columbia River Entertainment. 2002.

5. Beck Hansen. Midnight Vultures. Interscope Records. 1999.

6. Damon Dion Reed. Bibbles of Spuce-tame. CreateSpace. South Carolina. 2009.

7. Damon Dion Reed. Dreamstar. AuthorHouse. Indiana. 2007.

www.ingramcontent.com/pod-product-compliance
Lightning Source LLC
Chambersburg PA
CBHW030912180526
45163CB00004B/1800